LISTEN TO THIS

by Patty Whitehouse

CONTENTS

ROURKE CLASSROOM RESOURCES
The path to student success

Sound is All Around

Be very quiet for one minute and listen. Think of all the sounds you hear. Do you hear an airplane, or a car, or a person talking?

The world is full of sounds. They are all around you and come from many different places. Each sound is different. Sometimes you can tell what makes the sound by listening carefully.

The flute and piano make different sounds. You could tell them apart even if you were listening in another room.

Vibrations Make Sounds

When you play a guitar, you strum the strings. That makes the strings move quickly back and forth. The strings are **vibrating**.

When something vibrates, it may make a sound. You can see some vibrations, like guitar strings. You can feel some vibrations. You can feel the vibrations of a drum when you hit it. You can feel the vibrations of a passing train.

Sound Sources and Receivers

In order for you to hear a sound, two things need to happen. First, a sound must come from somewhere. That place or object is called the sound source.

The guitar is a sound source

Next, the sound must travel to a **sound receiver.** Ears are sound receivers. Animals and people hear sounds with their ears.

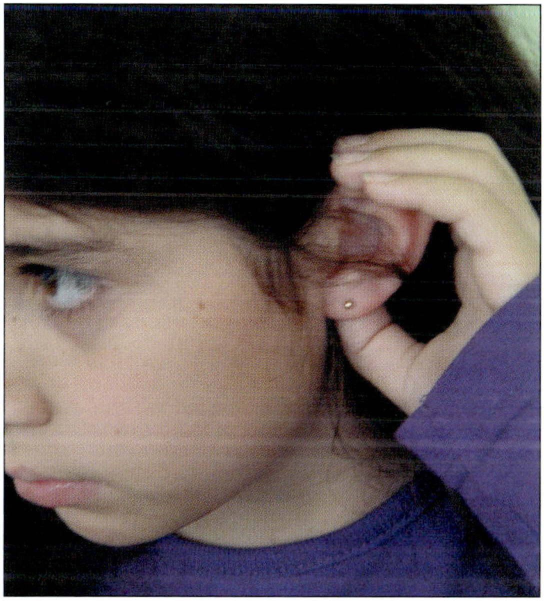

Sounds of Nature

You can classify sounds by their source. Some sounds are **natural sounds**. Their sounds source are living things. We can whistle, hum, sing, and speak. Animals make sounds. They can growl, grunt, whinny and moo.

There are other natural sounds besides animal sounds.You might be hearing some now. Wind, rain, and thunder are all natural sounds.

Storms make lots of natural sounds.

Man-made Sounds

Sounds that are not natural are man-made. Man-made sounds come from things people make.

The strings in the piano vibrate and make sounds when the keys are played.

For example, **instruments** can produce musical sounds. Cars, boats, and airplanes can produce noisy sounds.

The engine in this boat makes a noisy sound.

Pitch and Vibrations

Some sounds have a high **pitch**. Squeaks and whistles are high-pitched sounds. Flute music and many birdcalls are high-pitched sounds.

Some sounds have a low pitch. Grunts and growls are low-pitched sounds. The roar of a lion and the noise of a truck engine are low-pitched sounds.

Sound Source	High Pitch	Low Pitch
bird	X	
piano	both	
car engine		
airplane		
lion		
dog growling		

A sample student data sheet

Loud or Quiet

You can classify sounds by their **volume**. Volume describes how loud or quiet a sound is. A loud sound has a high volume. Screaming and airplane engine noise are high-volume sounds.

The volume of sound from the airplane engines is so high the man has to wear ear protection.

Some sounds are quiet. They have a low volume. A whisper is a low-volume sound. Listen carefully. Sounds are all around you.

GLOSSARY

instruments	man-made things that are designed to make musical sounds.
natural sounds	sounds made by something in nature.
pitch	how high or low a sound is.
sound receiver	thing that collects a sound
vibrating	moving back and forth.
volume	how loud or quiet a sound is.

INDEX